WASTE

ANNE O'DALY

BROWN BEAR BOOKS

Published by Brown Bear Books Ltd
4877 N. Circulo Bujia
Tucson, AZ 85718
USA

and

Studio G14, Regent Studios,
1 Thane Villas, London N7 7PH, UK

ISBN 978-1-78121-818-1 (library bound)
ISBN 978-1-78121-824-2 (paperback)

Library of Congress Cataloging-in-Publication Data available on request

Design: squareandcircus.co.uk
Design Manager: Keith Davis
Children's Publisher: Anne O'Daly

Manufactured in the United States of America
CPSIA compliance information: Batch#AG/5652

Picture Credits

The photographs in this book are used by permission and through the courtesy of:

iStock: Daisy-Daisy 6–7, AJ Watt 20–21; Shutterstock: Maxim Blinkov 10–11, Adam Gregor 16–17, Sony Herdiana 4–5, Natascha Kaukorat 18–19, KPixMining 14–15, RiskiBg 8–9, Skylines 12–13.

All other artwork and photography © Brown Bear Books.

t-top, r-right, l-left, c-center, b-bottom

Brown Bear Books has made every attempt to contact the copyright holder. If you have any information about omissions, please contact: licensing@brownbearbooks.co.uk

Websites

The website addresses in this book were valid at the time of going to press. However, it is possible that contents or addresses may change following publication of this book. No responsibility for any such changes can be accepted by the author or the publisher. Readers should be supervised when they access the Internet.

Words in **bold** appear in the Glossary on page 23.

CONTENTS

WHAT IS WASTE?

Rubbish has been dumped in this river. It makes the water dirty and harmful for people.

Waste is anything we throw away. It could be a banana skin. It could be a broken phone. Chemicals from a factory are waste. Almost everything we do makes waste.

TOO MUCH RUBBISH

The world's population is growing. All together, people produce 2 billion tons of waste each year. Most ends up buried in the land. Other trash goes into the rivers and oceans. Our world is turning into a massive rubbish dump.

Waste Comes from Different Places
Houses
Hospitals
Schools
Farms
Factories

HOUSEHOLD WASTE

What did you throw away this week? Food scraps? A magazine? An old T shirt? The average American throws out 1,360 pounds (620 kg) of trash a year. That's the weight of a grizzly bear!

WHAT'S IN THE TRASH?

Food waste is a big part of household waste. We waste about a third of the food we buy. Throwing food away is bad for the **environment**. It takes water, land, and **energy** to make food. All that is wasted when we throw food away.

Packaging

Many things we buy are wrapped in plastic packaging. Packaging makes up 65 percent of household waste.

Food waste can be collected. It can be turned into **compost.**

What Happens to Our Trash?

1. A garbage truck collects our trash. It crushes the trash so that it takes up less space.

2. The trash goes to a station. It is sorted to take out items that can be recycled.

3. Recycled materials go to a special plant. They can be turned into new things.

4. Most trash that can't be recycled goes to landfill. Some is burned to make energy.

LANDFILL

Most waste ends up in **landfill** sites. These are holes in the ground. Household trash is dumped in landfill. Waste from factories and farms goes to landfill, too.

BAD FOR THE PLANET

Landfills damage the environment. Some trash contains harmful chemicals. They leak into the soil and **groundwater**. Food waste rots. This gives off gases. The gases include **carbon dioxide** and **methane**. They are **greenhouse gases**. They heat the planet and cause **global warming**.

In some countries, people sort through rubbish. They look for materials they can sell.

Landfill and Pollution

Landfill sites make air, soil, and water polluted.

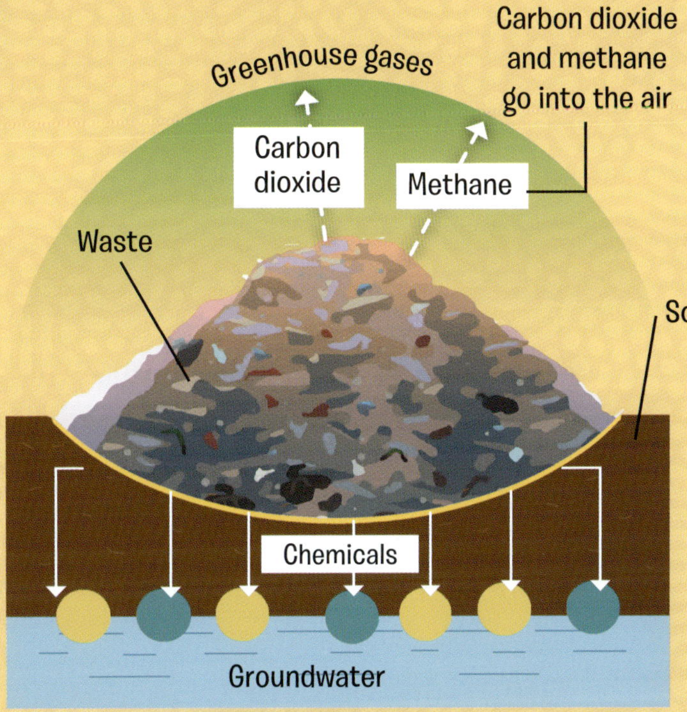

An Old Idea

The first landfill was in 3000 BCE in Crete. People buried their trash in deep holes. They covered the trash with dirt.

PLASTIC WASTE

Plastic is used to make lots of things. But plastic does not rot away. It just breaks into smaller pieces. Most plastic goes to landfill. A lot ends up in the oceans. Only a small amount gets recycled. Plastic waste pollutes the environment.

SINGLE-USE PLASTICS

Some plastics are made to be used once. They include drinking straws and grocery bags. Half of the plastic made each year is single-use. It makes 40 percent of plastic waste.

Plastic Roads

Inventors are using plastics to make roads! They take plastics that can't be recycled. The plastic is shredded. It is mixed with thick tar. Then the mixture is spread on the roads. Plastic roads last longer than normal roads.

Plastic garbage has been dumped on this beach. It gets washed into the ocean. It makes the beach dirty.

Different Plastics

Different plastics have different ingredients. Not all can be recycled. Here are some common plastics that can be recycled.

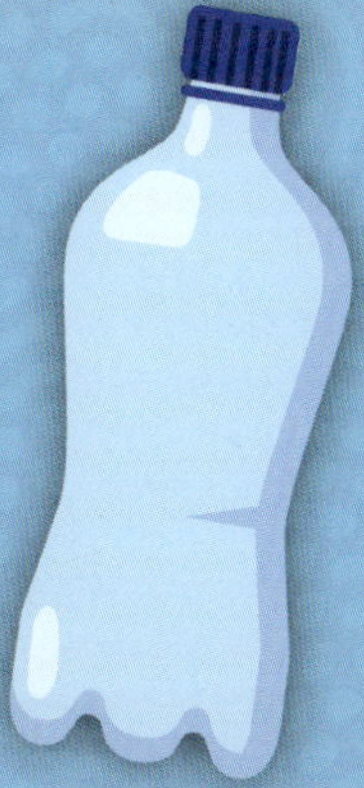

PET
Used to make water bottles.

PP
Used to make margarine tubs, straws, and bottle tops

HDPE
Used to make milk cartons, shampoo bottles, and toys

RECYCLING

Recycling turns trash into something new. Glass, paper, and metals can all be recycled. So can some plastics.

NEW FROM OLD

Making a product uses **natural resources**. Trees are cut down to make paper. Metals are dug from the ground. Recycling saves resources. It saves **energy**. It takes less energy to recycle than to make something new. And recycling keeps rubbish out of landfill.

Some houses use different bins. It makes it easier to recycle materials.

A TV can run for three hours on the energy saved by recycling one aluminum can.

Paper and cardboard can be recycled five to seven times..

How Much Energy Does Recycling Save

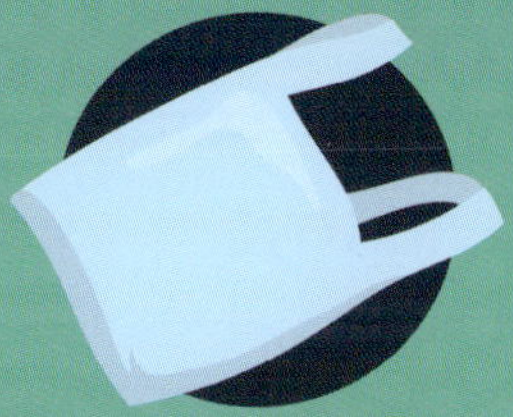

Recycling plastic saves 70 percent of the energy used to make new plastic.

Recycling paper and cardboard saves 40 percent.

Recycling glass saves 40 percent.

Recycling aluminum cans saves 95 percent.

Recycling steel cans saves 60 percent.

ELECTRONIC WASTE

Electronic waste is electronic devices that have been thrown away. Electronic waste is also called e-waste. Phones and computers are e-waste. So are kettles and toys.

Only 20 percent of e-waste is recycled properly. Dumping e-waste damages the environment. Chemicals from devices can leak into the soil.

PRECIOUS METALS

Phones and tablets contain rare metals. The metals are dug from the ground. More metals have to be **mined** to make new devices. Mining harms the environment. Metals in the Earth won't last forever. They will run out.

How Can I Help?

We can make less e-waste by keeping devices for longer. Can you buy a pre-owned device? Or have a broken one fixed?

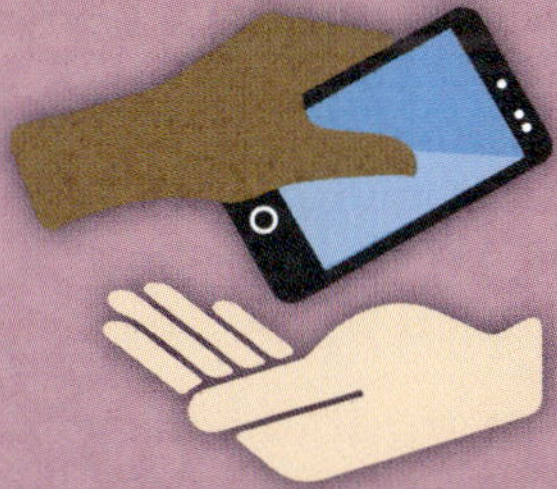

Give your old device to a friend instead of dumping it. There are also charities that collect old laptops and phones.

Recycle old devices. Valuable metals can be used again. Dangerous parts can be thrown away safely.

HAZARDOUS WASTE

Hazardous waste harms people. It damages the environment. Most hazardous waste is chemicals. They come from factories and farms. Chemicals also come from homes. Batteries and old paints contain chemicals.

KEEPING SAFE

Hazardous waste can't go into trash cans. It can't be recycled. It goes to special places. Batteries have chemicals that can leak. Hospitals make hazardous waste. Sharp needles are put in special bins.

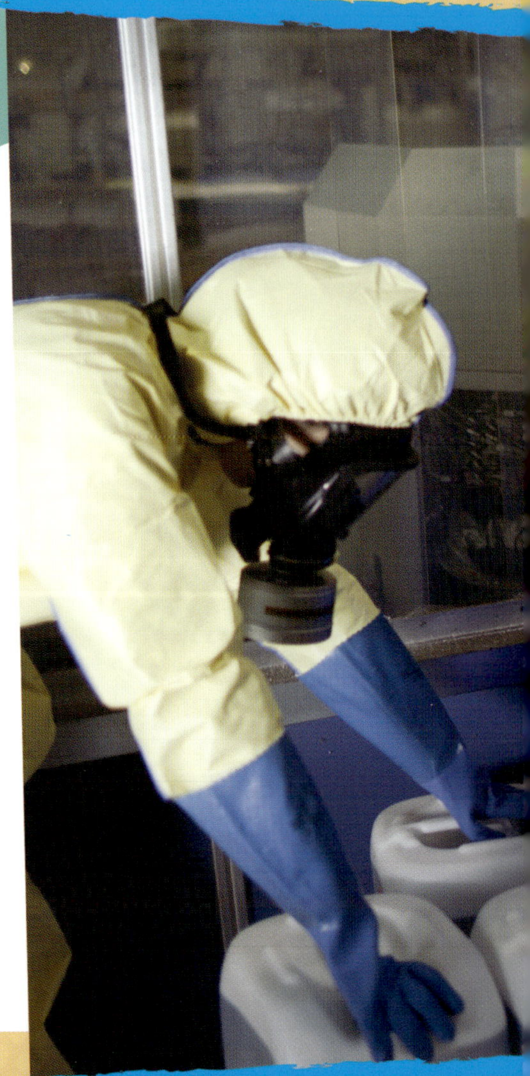

People working with hazardous waste wear protective clothing.

Hazard Symbols

Burns easily

Explosive

Very toxic

Irritates skin

Danger to the environment

ENERGY FROM WASTE

Trash can make energy. **Microbes** feed on food and farm waste. They break them down. This releases a gas. The gas is called **biogas**. It can be turned into fuel for cars and buses.

A biogas plant makes energy from waste.

BURNING TRASH

Trash can be burned to make energy. It is burned in a power plant. The heat makes steam. The steam makes huge turbines spin around. The motion makes electricity. Some people think the power plants make **pollution**. It is better to use less power and recycle.

Sources of Waste for Energy

Sewage (human poop)

Wood

Waste from crops

Waste from factories

Household waste

Animal waste

WHAT CAN WE DO?

Clothes are a big part of what we throw away. Why not arrange a clothes swap with your friends?

Waste is a big problem. It pollutes the planet. It harms the environment. We are running out of space to put it. But we can all do something to help.

REDUCE, REUSE, RECYCLE

We can reduce waste by buying less. Only buy things when you really need them. Use both sides of a piece of paper. Reuse the items you own. Give old toys and books away. Or sell them at a yard sale. Paper, glass, and metal can be recycled. Sort your trash before you throw it out.

Compost

We waste a third of the food we buy. One way to cut down food waste is to make compost.

1. Use a large plastic bin with a lid. Ask an adult to drill 8 to 10 holes in the bottom of the bin and the lid.

2. Add shredded newspaper until the bin is a quarter full. Put soil on top until the bin is half full.

3. Now you can add your scraps. Add layers of food scraps and paper.

4. Spray the compost with water. It should be damp but not wet. Every four to five days, stir it with a stick or spade.

5. The compost will be ready in one or two months. Mix it with soil in your yard to help plants grow.

You Can Compost:
grass clippings, fruit and vegetables, eggshells, coffee, tea bags, shredded newspaper, card egg cartons.

You Can't Compost:
meat, fish, bones, fats, cat and dog poop.

QUIZ

How much have you learned about waste? It's time to test your knowledge!

1. How much trash is produced each year?

a. 2 billion tons
b. 2 million tons
c. 2000 tons

2. Which gases are given off by landfill?

a. air and steam
b. methane and carbon dioxide
c. oxygen and carbon dioxide

3. Which of these are e-waste?

a. plastic straws
b. laptops
c. magazines about computers

4. What is biogas used for?

a. to put bubbles in fizzy drinks
b. to make burps louder
c. as fuel for vehicles

The answers are on page 24.

GLOSSARY

biogas gases given off by organic material including farm waste and food waste

carbon dioxide a gas that humans and animals breathe out, which is also produced when fossil fuels are burned

compost a mixture of natural materials that make soil healthier

energy the ability to do work

environment the natural world, especially the way it is affected by human activity

global warming an increase in Earth's temperature, mainly caused by human activity

groundwater water found under the ground

greenhouse gases gases that build up around Earth and trap heat

landfill an area where trash is buried

methane a greenhouse gas that is produced naturally by farm animals but also by human activity

microbes tiny living things that are too small to be seen by the eye

mined dug out of the earth

natural resources things that are found in nature and are useful to humans

pollution anything that makes Earth dirty

FIND OUT MORE

Books

Investigate Household Waste Pollution (Investigating Pollution). Sue Gagliardi, Wonder Publishing, 2022.

Waste and Recycling (Earth in Danger). Emily Kington, Hungry Tomato, 2022.

Your Planet Needs You! A Kids' Guide to Reducing Waste and Recycling. Philip Bunting, Bloomsbury Children's Books, 2022.

Websites

ducksters.com/science/environment/recycling_for_kids.php

kids.niehs.nih.gov/topics/reduce/index.htm

natgeokids.com/uk/kids-club/cool-kids/general-kids-club/tips-to-reduce-plastic-pollution/

INDEX

Answers: 1. a; 2. b; 3. b; 4. c.